YOUR MARE'S
FIRST FOAL

YOUR MARE'S
FIRST FOAL

JANE SKEPPER
BSc (Hons), AI Tech

KENILWORTH PRESS

Published in Great Britain by
Kenilworth Press, an imprint of Quiller Publishing Ltd

British Library Cataloguing in Publication Data
A catalogue record for this book is available from the British Library.

ISBN 1-872119-98-0
       978-1-872119-98-4

Layout and typesetting by Kenilworth Press
All drawings by Carole Vincer; those on pages 14, 19, 22, 25 and 45 are taken from *The BHS Veterinary Manual* by P. Stewart Hastie (also published by Kenilworth Press)
Photos by Piers Marson, Jane Skepper, Hannah Skepper, Poppy Skepper, Helen Davis (pages 9, 35, 65, 72 69, 81, 82), Charlotte Lowe (page 8, 54, 76), Trevor Meeks (page 30)

Printed in China

KENILWORTH PRESS
An imprint of Quiller Publishing Ltd
Wykey House, Wykey, Shrewsbury, SY4 1JA
tel: 01939 261616 fax: 01939 261606
e-mail: info@quillerbooks.com
website: www.kenilworthpress.co.uk

Acknowledgements

I would like to thank my mum, Hannah Skepper, whose experience is vast and incredibly practical, for passing on her invaluable knowledge over the years.

Thanks are also due to my vets, Iain Paton and Tony Lock, who provide us with superb veterinary support backed up by expert knowledge.

Finally, I'd like to thank my husband, Piers, for acting as my guinea-pig reader; and my first child, Ada, who has given me a new insight into motherhood.

Jane Skepper

Contents

Foreword

The Skepper family, owners of Heritage Coast Stud, have practical hands-on experience aplenty, going back at least three generations. Their unflappable common-sense approach has made them a pleasure to work with over the many years I have been a veterinary surgeon to the establishment.

I am delighted that at last these many years of practical experience, enhanced by Jane Skepper's up-to-the-minute techniques and scientific knowledge, are being committed to paper, thereby providing an invaluable handbook for successful horse breeding.

Iain Paton BVM&S, MRCVS
Paton & Lee Equine Veterinary Practice, Hoo Hall

Preface

Having stood competition stallions at stud for several years, it has become apparent to me that (professionals aside) the majority of individuals sending a mare to stud are first-time breeders.

Many horse owners want to produce a foal from their beloved mare who may have retired from competition; others simply like the idea of having a foal to bring on. Many of these first-time breeders come to the stud with limited understanding of the requirements for either sending a mare to stud, or actually producing a foal.

While there are plenty of well-written, high-level books written for the professional breeder, very little has been aimed at the novice. So, to begin with, I put together a hand-out for mare owners – and then it developed into something a little bigger: a straightforward, not overly scientific, guidebook on having your first foal. I hope very much that you find it helpful and that it fills in the gaps and answers most of your questions.

Happy breeding.

Jane Skepper
Heritage Coast Stud

CHAPTER ONE

Preparing your mare for stud

If you are considering breeding from your mare it is essential that she is physiologically in a good state to enable her firstly to get in foal, and secondly to stay in foal for the full gestation period and produce a healthy live foal.

Nutrition of the mare

Mares that are either too fat or poor in condition are difficult to get in foal. Many people overfeed their mares thinking that they should be on a rising plane of nutrition before going to stud. Instead you should aim for your mare to have good body condition, being neither too fat nor too thin.

The amount of food required by your mare depends on her breeding and individual metabolic rate.

- A Thoroughbred mare, for example, will usually require quite a high level of nutrition to maintain her condition through the winter, so that she is in good shape at the beginning of the breeding season.

- A pony-bred mare, on the other hand, will require good quality hay and grazing with virtually no supplementary nutrition before going to stud.

Monitoring your mare's cycles

You should monitor and keep a diary of your mare's cycles as soon as she starts to come into season in the spring. (Read Chapter 2 to find out what to look for.) This information will be very helpful to the stud and will enable you to be accurate about when to send your mare to stud and thereby reduce any unnecessary livery fees. Your mare's cycle is individual to her; some stay in season much longer than others. The cycle length information is vital for the stud, particularly if the mare has a very short season.

Book your mare in with the stud a week before she is due to come into season. This will give her time to settle into the new environment. If you have a mare with a foal at foot (this is a mare with a foal born in the current year) that you want to get in foal on her first heat, then send her to stud five days after foaling as she may come into season anything from five to ten days after the event.

Condition of your mare when sending to stud

- Make sure that your mare arrives at the stud looking tidy.

- She should have trimmed feet or, if she is shod, newly fitted front shoes and her hind shoes removed and feet trimmed.

- She should be in a suitable condition to be able to go out at grass during the day without rugs.

- If she has any vices, such as weaving or stable walking, the stud should be informed.

- If she suffers from any conditions such as sweet itch, COPD or any allergy, the stud should be informed.

- If your mare is on any medication when she is sent to stud, tell the stud and check with your vet that it will not have any detrimental effect on your mare getting in foal.

- It is a good idea to send your mare to stud with her own headcollar with her name on it. Some studs, especially those standing several stallions, may well have a large influx of mares during the breeding season, so anything that helps the staff to get to know each mare will be helpful.

- When leaving your mare at stud, take her passport and her swab test results.

- You should also give the stud some notes about your mare, including details about her temperament, her habits, her last farriery date, the

EXAMPLE OF COVERING LETTER TO LEAVE WITH STUD

Name of owner: Mrs P. G. Marson

Address: Heritage Coast Stud, Sudbourne, Woodbridge, Suffolk, IP12 2HD tel: ---------

Emergency: If no answer on above number, call mobile: -------

Fax: *Email*:

Name of veterinary practice:

Name of horse: Heritage Rusalka

Registration: Blue passport, Reg Anglo Arab Stud Book, Vol XV, No. 312

Brief description: Bay mare with white stripe, one white fetlock near hind, 16.1hh. She is an easy mare but can sometimes be funny about her ears being touched when putting a bridle on.

Last farriery: Trim all round 22/02/06

Last wormed: 22/02/06 with Eqvalan

Last vaccinated: 11/11/05 flu and tet

Feed: Ad lib hay, 4lb/2kg stud mix with chaff, morning and night

Last cycle: Started 3/3/06 ended 9/3/06

Other information: No vices, good with other horses.

date she was last wormed (and what with), her last cycle dates (start and finish), her normal feed and details of any supplements you have brought, the name of your vet and your contact details.

Health requirements

Your mare should be in good health when going to stud and should not have been in contact with any known infections. The stud will not appreciate you sending a mare that is not in complete health, and it will reduce her chance of getting in foal.

The stud will ask you to present 'swab certificates' to show that your mare is not carrying any venereal diseases. There are three main venereal diseases that are tested for by taking swabs prior to the mare going to stud.

Contagious equine metritis – CEM (*Tayorella equigenitali*)

Contagious equine metritis (CEM), caused by the *Tayorella equigenitali* organism, occurs widely in the non-Thoroughbred population, and to a limited extent in Thoroughbreds, in mainland Europe.

Klebsiella (*K. pneumoniae*)

There are many capsule types of *K. pneumoniae*, most of which do not cause venereal disease. However, types 1, 2 and 5 may be sexually transmitted. Therefore, when *K. pneumoniae* is identified in breeding stock, tests to determine the capsule type(s) present must be undertaken.

Pseudomonas (*P. aeruginosa*)

Not all strains of *P. aeruginosa* cause venereal disease but there is no reliable method to differentiate between the strains. Therefore, all positive results should be considered as potential infection.

Transmission of the above venereal diseases

Infection can be transmitted between horses in any of the following ways:

- direct transmission during mating

- direct transmission during teasing – an infected teaser can transmit disease to mares through contact with his genitalia

- indirect transmission during teasing – an infected teaser can transmit infected vulval discharge between mares through genital or naso-genital contact

- transmission to mares if semen used for AI comes from infected stallions or has been contaminated with bacteria during semen collection process

- indirect transmission via the hands and equipment of staff or veterinary surgeons who have handled the tail or genitalia of an infected horse

Prevention

The most important means of preventing infection are:

- establishing freedom from infection before commencing breeding activities

- checking that horses remain free from infection during breeding activities

- exercising strict hygiene measures during breeding activities

No vaccines against these above bacterial diseases are available.

Ensuring freedom from infection

Establishing freedom from infection before, and checking that horses remain free from infection during, breeding activities involves a veterinary surgeon taking samples ('swabs') from the genitalia of mares and stallions for testing ('culturing') in the laboratory. The laboratory will detect the presence of any bacteria.

- If the results are negative, the horse is free from infection and breeding activities may take place.

- If the results are positive, the horse is infected and must be treated, re-tested and cleared. The horse must not be used for breeding activities at this time.

No horse should be used for breeding activities until or unless all swab results are available and negative.

Types of swab

There are two types of swabs for mares:

- *Clitoral swab* – taken at any point during the reproductive cycle to demonstrate whether the clitoral fossa and sinuses are free from infection. In the case of pregnant mares, these swabs may be taken before or after foaling.

- *Endometrial swab* – taken during oestrus from the lining of the uterus

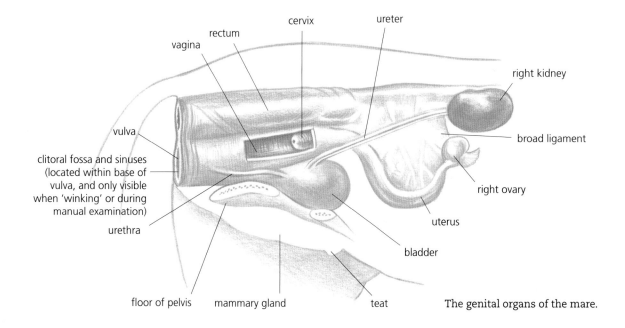

The genital organs of the mare.

via the open cervix to demonstrate whether the uterus is free from infection.

The laboratory results can take seven to ten days, so advance planning is required to get the swabs taken and the results back before going to the stallion.

Before sending your mare to be mated, you should request confirmation that the stallion is also disease-free in the current breeding season, and if possible view the up-to-date laboratory certificate for that stallion.

Mare owners should not accept semen for AI without obtaining evidence that the donor stallion was free from infection when the semen was collected. When importing semen, it should be supported by documentary evidence of freedom from infection with all three bacteria listed above.

An example of a mare's negative swab certificate.

In the UK, infections of these three diseases are notifiable under the Infectious Diseases of Horses Order 1987. The Codes of Practice published by the Horserace Betting Levy Board provide a complete outline of the requirements for venereally transmitted diseases in horses.

Another notifiable disease that is spread via breeding activities is Equine Viral Arteritis (EVA).

Equine Viral Arteritis (EVA)

EVA is caused by the equine arteritis virus. The virus occurs worldwide in Thoroughbred and non-Thoroughbred populations and is present in the United Kingdom, Ireland and every mainland European country. Amongst other things, it can cause abortion in mares, and is therefore significant.

Clinical signs of EVA

The variety and severity of clinical signs of EVA vary widely. Infection may be obvious but there may be no signs at all. Even when there are no signs, infection can still be transmitted and stallions might still 'shed' the virus, i.e. excrete it in their semen. These stallions are known as 'shedders' and pose a significant risk of disease transmission if undetected. In pregnant mares, abortion may occur. EVA may, occasionally, be fatal.

The main signs of EVA are fever, lethargy, depression, swelling of the lower legs, conjunctivitis (pink eye), swelling around the eye socket, nasal discharge, 'nettle rash' and swelling of the scrotum or mammary gland.

Transmission of EVA

Infection can be transmitted between horses in any of the following ways:

- direct transmission during mating
- direct or indirect transmission during teasing

- artificially inseminating mares with semen from infected stallions or which has been contaminated during semen collection or processing; the virus can survive in chilled and frozen semen and is not affected by the antibiotics added

- contact with aborted foetuses or other products of partuition

- via the respiratory route (e.g. droplets from coughing and snorting)

The 'shedder' stallion is a very important source of the virus. On infection, the virus localises in his accessory sex glands and will be shed in his semen for several weeks, months or years – possibly for life. The fertility of 'shedder' stallions is unaffected and they may show no clinical signs but they can infect mares during mating, or through insemination with their semen. These mares may, in turn, infect other horses via the respiratory route.

Prevention of EVA

The main ways of preventing EVA are to establish freedom from infection before breeding activities commence and to vaccinate stallions and teasers.

Freedom from infection can be determined by taking blood samples prior to the start of the breeding season.

Vaccination against EVA

Vaccination against EVA is particularly recommended for stallions. A vaccine called Artervac is available in the UK, manufactured by Fort Dodge.

Recommendations for prevention of EVA in domestic mares

In any breeding season, the safest option is to blood-test all mares after 1st January and within twenty-eight days before engaging in breeding activities. The mare should not be used until the results are available.

It is important to note that many studs vary on the policy regarding EVA testing of mares coming to stud. It is recommended that mare owners check that the stallion they are intending to use is either regularly vaccinated or has a recent blood-test showing that he is free from infection.

Equine Herpes Virus (EHV or equine rhinopneumonitis virus)

Equine herpes virus is a highly infectious disease which can cause abortion in mares. The virus can survive fourteen to forty-five days in the environment and is spread via the respiratory tract and from aborted foetuses, membranes and fluids.

Infected foals can pass the infection onto other healthy mares via respiratory systems.

Once a pregnant mare is infected she can abort from four months but usually between the eighth and eleventh months. Abortion may occur from two weeks after infection to several months later.

Symptoms of EHV

The symptoms are mild fever, coughing and a discharge from the nose.

Blood samples and swabs, taken from the throat of suspect cases, should be taken by a vet to confirm diagnosis.

Prevention against EHV

EHV is destroyed by heat and disinfectants, so it is advisable to steam-clean and disinfect your stables, using an approved product. The virus can survive several weeks in uncleaned stables.

Vaccinating mares with an approved vaccine can raise the level of protection but may not guarantee total immunity.

Other considerations before breeding from your mare

The mare contributes around fifty-five per cent of the foal's genes so it is important that your mare is good enough quality to breed from. You should consider particular areas such as:

- temperament

- conformation

- general health

- pedigree

- reproductive conformation and health

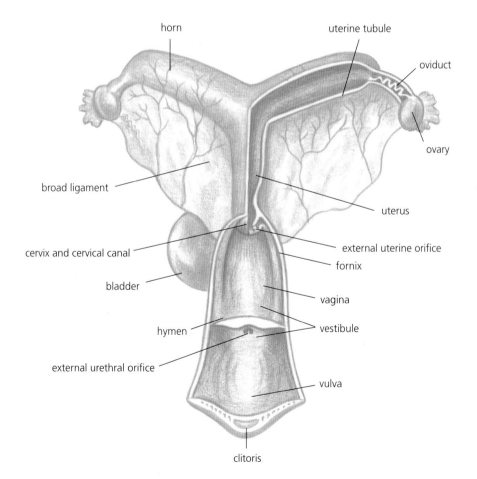

Dorsal view of the uterus.

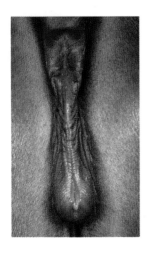

An example of a young mare with excellent breeding conformation.

Temperament – Your mare should have a sound temperament with no quirky behaviour. If she has vices such as crib-biting, weaving, windsucking and stable walking, she must be regarded as unsuitable to breed from.

Conformation and general health – Your mare's basic conformation and general health must be good enough to stand up to the physical aspects of breeding. Hoof and leg abnormalities can cause problems (for example, if your mare was born with one boxy foot or offset cannons, these may be passed on genetically), and pelvic injuries may result in foaling difficulties. Mares with parrot mouths or undershot jaws should be avoided as these features can be passed on.

Pedigree – A good pedigree is not essential but it helps when trying to breed quality.

Reproductive conformation and health – A vet can carry out internal examinations and scanning to evaluate the general health and condition of your mare's reproductive tract.

If your mare has a sunken anus, dirt can find its way into the reproductive tract. This type of mare is likely to require stitching once pregnancy has been diagnosed to reduce the risk of infection.

CHAPTER TWO

Understanding your mare's breeding cycle

Typically mares cycle, i.e. come into season, during the period of late winter to early autumn. This period of sexual activity is initiated by a series of hormonal events stimulated by increasing length of daylight.

Approximately every three weeks during this period the mare will exhibit receptivity to a stallion for some five to seven days. This spell of receptivity is known as **oestrus**. Between the three-week periods the ovarian activity subsides; this period is known as dioestrus. During autumn and early winter, in response to shortening daylight, ovarian activity is entirely absent; this is termed **anoestrus**.

Hormonal reasons for the states of breeding in the mare

Daylight periods are perceived by light receptors in the eye, which has an effect on the pineal gland, causing it to release a hormone known as melatonin. When melatonin levels increase, as the result of increased daylight hours, this results in the onset of oestrus, i.e. the breeding season. This explains why mares start coming into season in early spring, and (most) cease during winter.

There are several other hormones that are involved in bringing on the oestrus period:

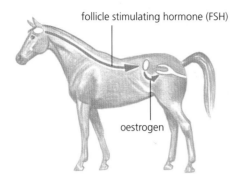

OESTRUS (0–5th day) – tail up, winking, cervix relaxed, tract moist

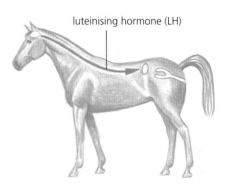

OVULATION (5th day) – tail up, winking, cervix relaxed, tract moist

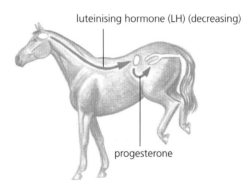

DIOESTRUS (7th day) – ears back, kicking, cervix closed, tract dry, progesterone dominant

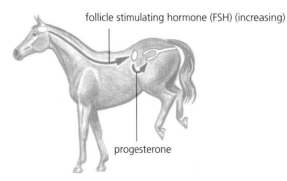

DIOESTRUS (13th day) – mid-cycle surge of follicle stimulating hormone, progesterone dominant

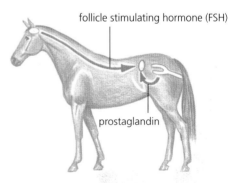

NEW OESTRUS starts (20th day) – follicle stimulating hormone and prostaglandin dominant

Oestrus cycle – hormonal stages in adult mare.
Note: Individual mares may vary by plus or minus a couple of days.

- Increased melatonin levels affect the hypothalamus part of the brain, which in turn results in the increased release (from the hypothalamus) of GnRH (gonadotrophin-releasing hormone).

- GnRH affects the anterior pituitary gland to induce the release of FSH (follicle stimulating hormone) and LH (luteinising hormone).

- FSH and LH have an effect on GnRH and gonadotrophins, and vice versa, during the oestrus cycle in a rhythmic fashion that repeats approximately every twenty-two days until the sequence is either

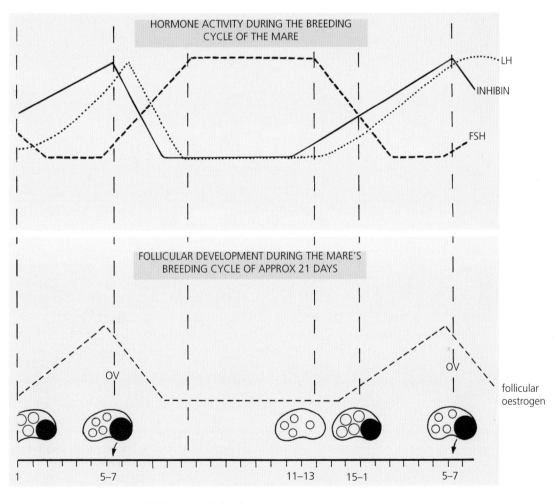

HORMONE ACTIVITY DURING THE BREEDING CYCLE OF THE MARE

FOLLICULAR DEVELOPMENT DURING THE MARE'S BREEDING CYCLE OF APPROX 21 DAYS

KEY: OV = ovulation   FSH = follicle stimulating hormone
LH = luteinising hormone   INHIBIN – inhibits FSH

Diagram showing the effects on the ovary of the hormones released by the brain, as a result of increased daylight hours.

KEY
OV = ovulation
FSH = follicle stimulating hormone
LH = luteinising hormone
CL = corpus luteum
GnRH = gonadotrophin-releasing hormone
-ve = negative effect
+ve = positive effect
inhibin – inhibits FSH
hypothalamus– area of the brain
pineal gland – gland in the brain

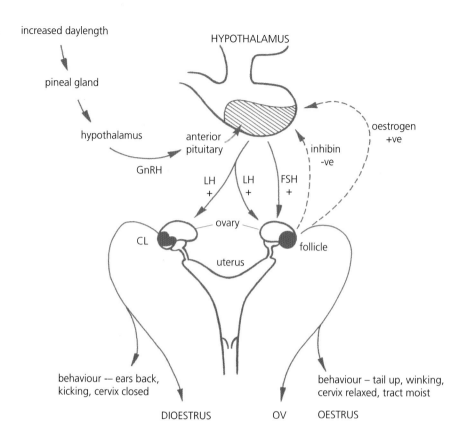

interrupted by pregnancy, a change in season (e.g. onset of winter) or pathological reasons.

At the same time as the hormones released from the brain are being produced, the ovary is also producing hormones in a similar pattern. The ovary goes through several stages in the twenty-two day cycle:

- Growth of follicles in the ovary results in the release of follicular oestrogen.

- Release of ovum on stimulation by LH, which results in the development of the corpus luteum (ovary post ovulation) and a reduction in follicular oestrogen.

- Growth of corpus luteum, resulting in increased progesterone production.

- In the event of no pregnancy the corpus luteum begins to decline, resulting in the reduction of progesterone.

Mare indicating oestrus.

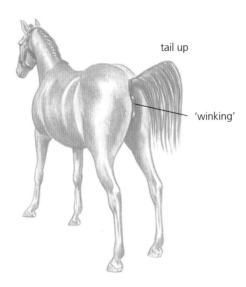

tail up

'winking'

- The cycle begins again with the growth of follicles and the release of oestrogen.

The oestrus cycle is divided up into two distinct phases – oestrus and dioestrus.

Oestrus – this is the period (0–5 days) when the mare exhibits signs of receptivity to the stallion. Usually this behaviour begins to wane about the time of ovulation, although some mares may carry on for a further one or two days post ovulation. Ovulation (this is when the egg is released from the ovary) coincides with the LH peak rise, resulting in increased follicular oestrogen and ovulation of one ovum (egg) from the ovary.

Dioestrus – this is the period when the mare is not receptive to the stallion. Her progesterone levels are high and the corpus luteum is in evidence (the ovary exhibits a corpus luteum, which is yellow in colour, where the egg was released from the ovary). This period is from day 7 to day 13. From the 13th to the 20th day the mare will appear in dioestrus (not receptive to the stallion) but it is during this period that the follicles in the ovary are developing, leading into the next oestrus phase (beginning around the 20th day). Hence the overall length of a mare's cycle, from oestrus to oestrus, is approximately 21 days.

The end of one oestrus phase and the beginning of another is marked by ovulation and is often used as a reference point for relating various events during the cycle. For management purposes at studs, ovulation (or the first date of covering) is taken as day 0. It is then possible to estimate that in 20–25 days from day 0, your mare should be coming back into oestrus unless she is in foal.

The ovary consists of many ova (eggs). During oestrus, follicles develop on the ovary, each containing an ovum. The follicles develop until the point when the most developed follicle releases its ovum into the fallopian tube (ovulation). Once ovulation has taken place the follicle develops into the corpus luteum.

CHAPTER THREE

Choosing a stallion

Type of stallion

Choosing a stallion is very much a matter of individual preference and what you are trying to breed. You may be breeding a pure-bred, e.g. Dartmoor pony, Arab horse or Thoroughbred, in which case you will need to find a stallion that is registered in the relevant stud book. If you are trying to breed a competition horse, such as a show jumper, eventer, show pony, dressage horse, endurance horse, etc., you will be looking for a stallion that will have 'form' (competition results) in your chosen sphere.

There are various independent bodies – for example, in the UK, NaStA (National Stallions Association), SHB (GB) (Sporthorse Stud Book of GB), AES (Anglo European Studbook), BWBS (British Warmblood Society), SPSS (Sports Pony Studbook Society) – that operate stallion gradings.

The grading procedures vary with each stud book, but they all assess conformation and an individual's suitability as a breeding stallion. Some gradings are extensive and include ridden examinations, loose-jumping tests, ridden jumping and a full five-stage vetting.

These gradings are an excellent guide for the mare owner, confirming the quality of a chosen stallion. To find out more about a particular stud book's grading requirements, you should contact the stud book/society directly.

Many competition stallions are able to compete and stand at stud effectively through the introduction of artificial insemination. This enables a stallion to be competing regularly while his stud duties are also being carried out. The use of artificial insemination (AI) is widespread in Europe and is becoming more popular year on year in the UK. The main advantage for the mare owner is that the choice of stallion is vast, as you can use an overseas stallion without the expense (or worry) of sending your mare to stud abroad.

Viewing a stallion

When you have drawn up a short-list of stallions that might suit your mare, make appointments to view them in the autumn. If you are breeding a Thoroughbred, it is sensible to view the stallions as early as possible. Some are restricted on the amount of mares they will accept, so it may be important to get your application in early to avoid disappointment. If you are breeding a non-Thoroughbred, there is usually not so much urgency as they will often not have a restriction on the number of mares they will accept.

When viewing the stallion there are several aspects that you should examine thoroughly:

Conformation (especially important)

- **Head and eye** –look for an honest head with a kind eye

- **Shoulder** – a good quality shoulder is essential

- **Forearm** – is he 'over at the knee' or 'back at the knee'? ('back at the knee' should be avoided)

- **Overall length** – not too short or long in the back

- **Feet** – look carefully at the shape of the feet. Good feet are essential. Small feet should be avoided

- **Bone** – check the amount of bone. Do not breed from a horse that is light of bone unless he has produced proven quality youngstock

If you are not confident about analysing a horse's conformation seek advice from someone with experience. Stallions that have been graded for a stud book will have had their conformation evaluated by professionals, so the fact that the stallion is graded should provide you with some confidence that he is suitable for use as a stallion.

Temperament

Most stallions that are competing will have a good temperament. However, when viewing a stallion you can get a good impression of his overall demeanour.

Size

The size of the stallion will be directly related to its breeding and its discipline, i.e. dressage stallions tend to be tall and have plenty of bone, whereas Arab stallions will be fine and not much bigger than 15.2 hh. Some breeds are divided into different classes according to their height, and this should be considered if you are hoping to produce a horse or

The graded Anglo Arab stallion, Heritage Orion, who has a proven competition record.

A dressage stallion showing excellent movement.

pony for a particular section of a breed. For example, Welsh ponies are divided into sections A, B, C and D, with A being the smallest and D the largest.

If you are trying to breed a show pony or horse, showing classes are divided according to height, so again height should be considered when choosing your stallion.

It is often sensible to choose a smaller stallion or one that has a tendency to produce smaller foals when breeding from a maiden mare (first-time mother). This will reduce the chances of her having a very large foal and thereby risking foaling complications.

Movement

Ask to see the stallion walked up and down and then trotted up and down. Check that he moves straight and has good elevation to his steps.

Colour

The genetics of colour in horses is complicated, with certain colours, such as pure black and palomino, being very difficult to produce. Using a coloured stallion on a plain mare (or vice versa) does not necessarily result in a coloured offspring. Colours that are generally recessive

			Chestnut stallion
		Grey stallion	
			Chestnut mare
	Bay stallion (A)		
			Bay stallion
		Brown mare	
			Bay mare
Chestnut stallion			
			Grey stallion
		Grey stallion	
			Chestnut mare
	Brown mare (A)		
			Brown stallion
		Brown mare	
			Brown mare

The brown mare (A) has bred 4 brown mares and one chestnut colt when mated with the same bay stallion (A)

A pedigree showing offspring colours and indicating how difficult it can be to predict the next generation accurately.

(weaker and less likely to be carried through) are chestnut and grey.

It is important to note that some colour genes are sex linked. This means that they are carried on the Y chromosome (male sperm) and will be dominant if a male foal is produced.

Youngstock

Usually, there will be some youngstock of the stallion on site, as most studs are breeding as well as standing stallions. Ask if you can view any of his youngstock. This is an excellent way to see what traits the stallion may to be passing on. This is not often available with Thoroughbred stallions, but you can view youngstock at the various foal and yearling sales in the preceding autumn to sending your mare to stud.

Finally, will the stallion complement your mare? No horse is perfect, and your mare will have both faults and strengths. Ideally, the stallion should offer compensatory strengths to your mare's weaknesses.

Checklist

When sending your mare to stud, you want to be sure that not only will she be looked after well but also that the stallion is in good condition for the breeding season to provide the best rates of fertility.

When viewing the stallion, here are some specific questions you should ask:

- **Fertility** – what was last year's fertility rate, i.e. what percentage of visiting mares conceived? You can further clarify this by asking whether most mares 'took' on their first cycle.

- **Vaccinations** – is the stallion vaccinated against EVA (equine viral arteritis)? This is very important as stallions can be 'shedders' of this disease. This means that they can pass it onto mares via their semen without showing any symptoms of having EVA. EVA in mares can cause abortion and infected mares can get ill and then further pass it on via their respiratory system (sneezing).

- **Swabs** – the stud will usually require you to produce swab certificates for your mare showing that she is free of venereal diseases. Equally, you should request evidence that the stallion you are sending your mare to is free of any infection. Stallions should be swabbed at the beginning of every breeding season prior to mating any mares.

Nominations

When you have chosen the stallion for your mare, ensure that you get a nomination agreement in writing before making any commitment. Nominations are commonplace in the Thoroughbred industry but rare in the non-Thoroughbred breeding world, so it is advisable to take your own. There are a number of potential areas of dispute between stud and mare owner, and if you do not have written terms and conditions to refer to it is extremely difficult to take action if something goes wrong.

While the bloodstock industry is not exactly renowned for its use of written contracts, this is one area in which there is a tried and tested agreement that is widely used. The Thoroughbred Breeders Association

Standard Terms cover all the key issues that both parties need to be aware of. If disputes do arise, they are resolved by arbitration rather than going through the courts. Most breed societies can offer nomination agreements.

Not surprisingly perhaps, nomination fees are the biggest source of problems. These may range from as little as, say, £100 for a pony to well over £100,000 for a top Thoroughbred stallion. Whatever the amount agreed, both sides need to be quite clear about when the fee is payable and under what circumstances it might be returned.

Terms will vary from one stud to another. Some may require full or part-payment up front, or payment on collection of the mare when scanned in foal; others will work on a 'no foal, no fee' basis, or request payment on 1 October if the mare is in foal on that date.

Other issues that need to be considered include:

- **Mating** – how many times must you make the mare available for covering? Some studs require that the mare is mated on three cycles if she does not hold (get pregnant) on the first cycle. A late expected date of birth can be very significant. In the Thoroughbred industry the lateness of a foal can affect its sale price.

  All horses become a year older on **1st January**, disregarding their date of birth. Therefore the later a foal is born, the further behind it will be when competing against its peers. This has a greater effect when the offspring are under the age of five years; after that there is generally very little difference between a horse born early in the year and one born later as they are now both fully mature.

- **Refusal of service** – under certain conditions, the stud/stallion owner may reserve the right to refuse to allow the stallion to cover the mare. Be sure that you are aware what those conditions are – and where you stand in terms of advance-paid fees if that does happen to you.

- **Substitution** – if your mare dies before being served, or is unfit for service for some other reason, will you be able to substitute another mare?

33

- **Death or sale of the mare** – where do you stand on fees if your mare dies after being served by the stallion, but before the foal is born. Alternatively, who pays the nomination if she is sold in foal? Normally, the original purchaser of the nomination is the debtor.

- **Disposal of the stallion** – what compensation (if any) will be offered if the stallion is sold or otherwise disposed of by the owner before serving the mare? This is particularly relevant to mare owners claiming a 'free return' in successive years.

Many stud/stallion owners have their own agreements, covering all of the issues raised above. Be sure you understand the terms of the contract. If no agreement is offered, don't be afraid to insist on something in writing.

CHAPTER FOUR

Getting your mare in foal

You have now booked your mare into your chosen stallion. The stud may offer the opportunity to use the stallion by natural covering and/or by artificial insemination (AI). There are various advantages and disadvantages to each method and you should evaluate these before deciding which to use.

Natural covering/service

If choosing natural covering for your mare, you can expect to get the best rates of conception by this method.

If you decide to send your mare away to stud, you obviously need to satisfy yourself that she will be well looked after, kept safe, and given every chance to get in foal. You will be relying on the stud management of the establishment to make sure that all these things happen, and you need to be convinced that this is so.

Modern stud management means that busy stallions can now cover a greater number of mares than ever before, since coverings when the mare is unlikely to ovulate need not be made. The mares can be scanned regularly to monitor the ovary, and when they are about to ovulate they will be put to the stallion. Many mares are now only covered once per cycle. This also reduces the chance of infection.

Where stallions are less pressured, which is the case for most non-Thoroughbred stallions, a less hi-tech approach can be operated simply because it is much cheaper for mare owners. Here mares are teased every other day by a stallion until they come into season. The mare will then be covered on the third day, if she is willing, and again on the fifth. On the seventh day, the mare is generally no longer receptive – but if she is still receptive she is scanned to see what stage the ovary is at and whether ovulation is imminent. If not, the vet will administer hormones to encourage her to ovulate and then she can be covered again.

When the mare is covered naturally, she is covered in hand. This means that the stallion and the mare are held during the mating. The mare wears a neck guard to protect her neck from being bitten too badly, and she will also wear 'covering boots' (thick felt boots on the hind feet) to protect the stallion from being kicked. Occasionally a mare may be **twitched** if she is very difficult to hold for the stallion (usually only maiden mares).

twitched – this is when a man-made tool called a 'twitch' is placed tightly around the muzzle of the mare, resulting in endorphins being released that relax and calm the mare, making her easier to hold.

Artificial insemination with chilled semen

Sperm is collected from the stallion and chilled. Chilled semen will usually last around twenty-four hours once inseminated, but this will vary from stallion to stallion. Insemination needs to occur within a twenty-four hour period of the mare ovulating. The veterinary work is quite substantial, although less so than with frozen semen.

The mare needs to be scanned probably twice in the heat period to try to get the semen delivered as close to ovulation as possible. She may need treatment to try and regulate her cycle. She may need further treatment to try to make her ovulate once she has been inseminated. Then she will need scanning twenty-four hours after insemination to make sure that she has ovulated. If she has not, then more semen will be needed, and must be despatched for next day delivery and the process begins again.

Many people might imagine that using AI, as opposed to sending their mare away to stud, is a cheaper way of getting their mare in foal. However, this is not the case and very often is more expensive due to the additional veterinary work that is required. It does, though, have the advantages that there is no risk to your mare being injured while being covered naturally, and the risk of disease is reduced as the whole process is more clinical.

Your mare can be inseminated at your own location or you can send her to an AI centre. The advantage of using an AI centre is that they will probably be able to give you a package price for the insemination, monitoring and scanning and so control the veterinary fees. They are professionals at managing mares and so will have good success rates with AI. They will also have the facilities to examine the semen on arrival to be sure that it is in the best condition before being inseminated.

The main advantage of using AI is that you can use a stallion located anywhere rather than being controlled in your choice by the distance that your mare will have to travel.

Artificial insemination with frozen semen

This method is technically demanding, partly since thawing the sperm needs precision and care, but even more so because the sperm does not live long once it is thawed, and the mare needs to be inseminated within five or six hours of ovulation. Thus it is really not possible for the one-horse owner to have this done at home. The mare needs to go to an AI centre or to a veterinary hospital, where they are using the appropriate technology regularly and have veterinary and support staff on duty at unsocial hours. Mares may need examining up to three times in each twenty-four hours to make sure that insemination happens close enough to ovulation to be successful, and at least one of these examinations will be at night. This is not cheap at all, and the cost of a cycle of treatment and insemination will cost as much as several weeks' keep at a stud.

The conception rates are less good than with chilled semen. Some stallions simply 'do not freeze' and so their semen is not available by this method. The great advantage of this technology is that 'the world is your oyster' – you can choose a stallion from anywhere around the globe, and since the semen is not perishable whilst frozen, it can be got into place well in advance of being needed.

Pregnancy diagnosis

Once your mare has been inseminated or mated you will need to wait until fourteen days from ovulation (or first date of covering where mares are not scanned for ovulation) to be able to determine whether or not she is in foal.

Pregnancy diagnosis can be carried out by three methods: scanning, blood-testing and manual examination.

Scanning

After mating, your mare can be scanned for pregnancy from, at the earliest, fourteen days from the first date of covering. Scanning for early pregnancy is highly specialised work and will be carried out by a vet, either on site at the stud or at a veterinary practice.

Usually, the stud will have a set of stocks into which the mare is led and held while the vet scans her. The stocks help to keep the mare still and ensure the safety of the vet and the mare during the scan.

Scanning is carried out by the vet using an ultrasonograph. The vet inserts the head of the ultrasonograph rectally, in order to be able to scan the mare's uterus. The vet will manoeuvre the head to scan both the left

Mobile ultrasonograph scanner.
Before scanning, the vet will evacuate all faecal matter from the mare's rectal channel. He/she will then insert the scanner head into the rectal channel and scan the mare's uterus and uterine horns to check for pregnancy.
As the scanner head moves around, you can view the scanning image on screen.

and right horns of the uterus to examine for a pregnancy. He will also be checking that there is not more than one pregnancy.

Your mare should be scanned twice. Once at this early stage, recommended between 16–19 days from first service date, and then subsequently at 24–30 days from the first service date.

The first scan is to check for pregnancy and also for any twins. If twins are present it is usually possible for the vet to intervene at this stage and 'pinch' one of the pregnancies. (This will not be possible if the pregnancies are close together in the same uterine horn.)

Routinely, the second scan should be carried out between 24–30 days after the first date of service to evaluate the pregnancy development. At this stage, the presence of a heartbeat can be detected, and it is possible to recheck for twins. If twins were present at the first scan and one was 'pinched', the second scan can confirm that only one pregnancy remains and that it is developing well.

Some mares may have lymphatic cysts, which can look very much like a pregnancy, but unlike a pregnancy they do not grow and, more importantly, won't develop a heartbeat. Therefore, a second scan will enable a clear differentiation between a pregnancy and a cyst.

Twins

If twins are detected close together in the same uterine horn, regular monitoring (usually at forty-eight hour intervals) is required by scanning. Usually, in this case, one of the pregnancies will be more dominant, and the second pregnancy will be reabsorbed, leaving one to develop. If, by twenty-four days, both pregnancies are still present, it will be necessary to abort them both and mate your mare again, when she comes back into season approximately ten days later. If twin pregnancies are still *in situ*, the latest they can be aborted is forty days from first date of service, by a vet using prostaglandin by injection.

• The statistics for the success of equine twins are not good, which explains why it is necessary to check for twins and act appropriately. Only 9 per cent of mares will carry twins to term, and of this 9 per cent some 64 per cent are born dead; in 21 per cent of cases one is alive, and in 14.5 per cent of cases both may be alive but many will die in the first week of life.

- Most mares will abort twins in late pregnancy.

- Mares that carry twins (either to term or to late pregnancy) are more likely to retain their placenta, and will be harder to get pregnant in the same season.

- Therefore, allowing twins to go to term is not only risky but also economically expensive, as you are liable to incur veterinary fees for the treatment of your mare and also the loss of a breeding season.

- Some mares have a genetic predisposition to conceive twins. If a mare has already shown a likelihood of twinning she must be monitored very carefully. Twinning is also more likely in certain breeds of horses, such as Thoroughbreds and Warmbloods.

Blood-tests

At forty days, the development of the endometrial cups (formed by the placenta embedding into the uterine wall) begins and the mare does not come back into season. The endometrial cups secrete ECG (equine chorionic gonadotropin) and this results in the formation of the corpus luteum, which in turn secretes progesterone. Progesterone is the hormone that maintains pregnancy and prevents the mare from having seasons.

ECG levels can be measured in the blood from forty days, which will indicate whether your mare is pregnant. It is important to note that many mares can reabsorb a pregnancy at this early stage and ECG will still be detected in the blood. It is wise, therefore, to follow this up with a blood-test, with further tests at sixty days to detect the rise in progesterone, which confirms that the pregnancy is being maintained.

For mares that have a history of early loss of pregnancy, supplementary progesterone therapy may be recommended for the first hundred days of pregnancy. Your vet will advise on how to proceed. The majority of mares, however, do not need this supplementation.

Manual diagnosis

Manual diagnosis of pregnancy can be carried out by a vet during the second stage of pregnancy. This requires the vet to carry out a rectal examination to 'feel' for the physical presence of a foal.

This examination is usually carried out after 1st October (for a mare mated between mid-February and mid-July) to enable the stud to confirm that a mare is still pregnant and should receive the required enhanced nutrition and care for the latter stage of pregnancy.

In the UK many stallion nominations will carry the phrase '1st October Terms'. This means that a veterinary certificate of pregnancy or no pregnancy is required in order to exempt the owner from the final stud fee payment (if the mare is not in foal), which otherwise is due at 1st October.

CHAPTER FIVE

Caring for your pregnant mare

Now that your mare has been confirmed pregnant, it is important to be sure to keep her in good condition to maintain a healthy pregnancy.

Nutrition

For the first eight months of the pregnancy, the nutritional requirements in the pregnant mare are no different to a non-pregnant mare. The mare should be healthy and maintained with good body condition and not fed excessively.

- You should determine the body condition score (BCS) of your mare at the early stage of pregnancy. BCS is based on a scale of 1–9, with 1 being emaciated and 9 being obese. For a healthy pregnancy a mare should have a BCS of 5 or 6 in early to mid-pregnancy and 6–7 in late pregnancy. Mares should gain weight in the last three months of pregnancy in direct proportion to increases in foetal weight.

- Weight gain can be measured monthly using a weight tape, and your mare's BCS checked at the same time. It should be noted that a mare can gain weight without her BCS changing. When examining for BCS, mares tend to lay fat down on the neck, withers, back, the dock and ribs. BCS is determined by visually examining the mare.

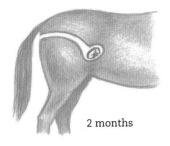

2 months

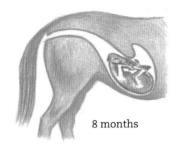

8 months

Development of the
foetus during the
gestation period.

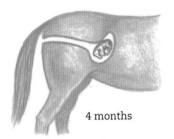

4 months

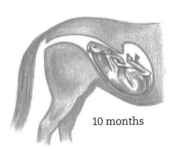

10 months

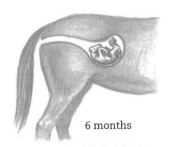

6 months

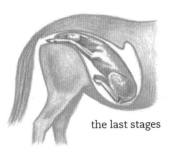

the last stages

- In the late stage of pregnancy (last three months) the energy and protein requirement of the mare will increase as the demands of the foetus increase. Generally, mares are fed 1–1.5kg of grain/100kg of bodyweight, with a protein content of 14 per cent. However, this may vary for different breeds, e.g. a native pony breed may be able to receive the increased energy and protein from better quality hay, whereas a Thoroughbred may well require more concentrate feeding to maintain a good BCS in late pregnancy. Overfeeding at this stage in some breeds may result in laminitis, which can be detrimental to both mare and foal.

- Most importantly, any changes to diet must be made gradually.

Mineral supplements

Pregnant mares should be supplemented with **calcium**, **phosphorus** and **salt** throughout pregnancy. If the land where the mare is being grazed is known to be deficient in any minerals, you should supplement these as well. There are now some very good supplements on the market designed specifically for the pregnant mare.

Other minerals that should be present in a supplement for the pregnant mare are: **iron, manganese, zinc, copper, cobalt, iodine** and **selenium**.

To summarise, good nutritional management of the pregnant mare includes feeding a well-balanced diet with an increasing plane of nutrition in the last three months of pregnancy. This policy will support the development of the foetus to term and ensure that the mare is in good condition for lactation.

- The most common errors are overfeeding the mare in pregnancy, and underfeeding the lactating mare, when nutritional requirements are at their highest.

A young mare at the ten-month stage of pregnancy.

Work

Regular exercise in healthy, pregnant mares is beneficial to the mare's circulatory system. Some competing mares may show improved performance in early pregnancy due to the increased progesterone levels. You can continue exercising your mare until around seven months into her pregnancy. However, this varies from mare to mare, depending on how relaxed she remains and how big she gets. A maiden mare should be able to carry on with work longer than a mare who has bred several foals and may well get larger earlier on.

Stressful exercise in mid- to late pregnancy is not advised. You should cease exercising when your mare appears to blow with light work or seems to get stressed or sweats overly.

It is fine to reduce your mare's exercise once she is tested in foal but she should be allowed to go out at grass as much as possible to exercise herself and keep in good condition.

Some racehorses are kept in training and race well when in foal. If a race mare is tested in foal in March she can continue training and racing until September. Obviously, when an in-foal mare is being exercised to such a level her nutrition and management must be of the highest quality (professional advice should be sought).

In some mares, exercise causes 'windsucking' via the vulval seal. This can happen if the perineum muscles relax, causing the mare to take air into her vagina. This can lead to an infection in the reproductive tract, and so it is advisable to consult with your vet. Your vet may well advise a small operation called Caslick's surgery. This is a simple procedure involving a few stitches in the vulva, once she has been confirmed pregnant. The only problem is that the mare has to be 'opened' (stitches and skin cut open) at the time of foaling.

Drawing showing how a mare's vulva is stitched in Caslick's surgery.

Turning out

Once your mare is in foal it is perfectly acceptable for her to be turned out with other horses. It is sensible, though, to turn her out with other mares, rather than with geldings or in a mixed group who might get boisterous. Ideally, in-foal mares spend the summer out together, and are then stabled at nights and out during the day in winter.

Vaccinations

If your mare is already vaccinated for equine influenza then continue with your normal annual booster. If your mare is going away to stud, the stud may have specific vaccination requirements that you will have to adhere to.

Worming rotation

DEC/JAN	FEB/MAR	APR	MAY/JUN/JUL/AUG	SEP/OCT	NOV
Treat for bots; killing the bots inside the horse's stomach before release via the horse's dung.	Encysted small redworm hide in the gut wall. If untreated will emerge in spring.	Tapeworm infection of horses.	Grazing period. Routine roundworm control including large and small redworm.	Summer grazing can result in greater exposure of horses to tapeworm.	The small redworm larvae ingested during the summer grazing embed in the gut wall.
Treatment: Moxidectin or Ivermectin.	Treatment: Single dose of Moxidectin or Ivermectin or 5-day course of Fenbendazole.	Treatment: Single dose of Praziquantel-based wormer or double dose of Pyrantel-based wormer.	Treatment: Ivermectin every 8–10 weeks, or Pyrantel every 4–8 weeks, or Fenbendazole every 6–8 weeks, or Mebendazole every 6 weeks, or Moxidectin every 13 weeks.	Treatment: Single dose of Praziquantel-based wormer or double dose of a Pyrantel-based wormer.	Treatment: Single dose of Moxidectin or Ivermectin, or 5-day course of Fenbendazole.

NOTES
Moxidectin is not advisable for use with foals. Consult your veterinary surgeon regarding the types of wormers to use on your foal.
Pyrantel wormers only control adult worms.
Mebendazole is not licensed to control the migratory larvae of large redworm.
All brands of wormers have their main constituent listed on the packaging to enable you to identify which wormer to use when.

It is usually recommended that you vaccinate your mare against tetanus toxoid and equine influenza six to four weeks prior to foaling, not only to cover your mare but also to provide protection to the newborn foal from these infections via the mare's colostrum.

If your mare's annual booster is due in this time then proceed with it as usual; no further vaccination will be required. If the annual booster is due earlier, then vaccinate with tetanus prior to foaling as described. If the annual booster is due post foaling then bring it forward to four to six weeks prior to the foaling due date.

Your vet may advise other, not so common, vaccinations if incidence of infection has been notified in your area, such as strangles.

Worming

- Mares should be wormed on a regular basis using products approved for use in the pregnant mare, such as benzimidazoles, fenbendazole, pyrantel embonate and avermectins. (The main constituents are listed on the packaging. If in doubt, ask your vet.)

- It is particularly important to worm your mare before foaling to make sure the foal is not exposed to a high level of worms, which can be passed on through milk. Also, foals may eat dung, through which worm eggs can be passed on.

- A regular worming programme should be maintained during the mare's pregnancy with the wormer types being rotated to avoid a build-up of worm resistance.

CHAPTER SIX

Preparing to foal down

The foaling environment

If you are intending to foal your mare in a different location, e.g. a larger foaling box, new isolated yard or different stud, you should move her into the new environment at least a month before her due date. This will minimise any impact of stress on the mare and allow her to build up immunity to any different bugs in the new location.

Ideally, house your mare in a large foaling box (usually double the size of a standard box) which has previously been cleaned out and disinfected. The box should be bedded down deeply, to protect both the mare and foal during foaling. If possible, the foaling box should have the facility for an infra-red lamp to be used, and windows to allow monitoring throughout foaling without disturbing the mare. Good ventilation is essential but, equally, it is important that top doors can be shut to keep the stable warm if foaling early in the year.

If you re-use the foaling box for another mare, the box must be completely cleaned out, disinfected and left empty for at least two days before re-bedding for the next foaling.

Vaccinations

As described in Chapter 5, the mare should be vaccinated for tetanus toxoid and equine influenza one month prior to her due date to ensure that the foal is protected at birth and for the first four months of its life.

Signs of imminent foaling

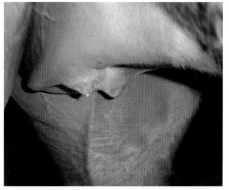

- **In the last month** of pregnancy, there is an increase in mammary development (the 'udder' or 'bag' increases in size). The secretion of fluid into the udder begins around six weeks prior to foaling. The consistency and colour of the secretion goes through a series of changes, which cannot be seen. The first sign of secretion that can be seen is the wax-like 'corks' on the end of the teat.

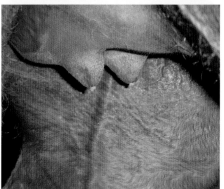

- **The week before** foaling, the mare will appear more 'floppy' around the tail end – the muscles either side of the tail in the hindquarters begin to slacken, resulting in them feeling loose and lacking in tone. This is particularly noticed if the mare is regularly checked by gentle hand pressure or when grooming daily during the month prior to foaling.

- **'Waxing up'** – a small amount of secretion can be seen hanging from the teat opening, resembling a cork. The 'cork' looks like candle-wax, hence the term 'waxing up'. Waxing up usually takes place within twenty-four hours of foaling but this varies from mare to mare and may occur several days before foaling and, in a few cases,

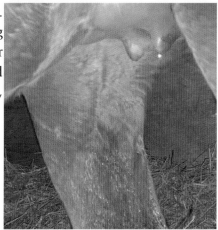

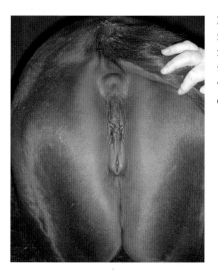

Mare beginning to appear more 'floppy' before foaling. The creases in the upper half of the vulva start to straighten out.

The 'waxing up' process. The above images show the development of the cork-like secretion on the end of the teat. This mare foaled ten hours after the last photo was taken.

49

not at all. When liquefaction of wax occurs (i.e. it turns to liquid) foaling usually takes place within twenty-four hours.

- **The final changes** – the vulva will relax and the creases in the upper half will appear to straighten out in the last twelve hours before foaling.

Supplies required for foaling down

Listed below are some basic supplies that you should have in readiness for foaling your mare:

- Headcollar and lead rope – for mare.

- Tail bandage – for mare during foaling, to keep the tail out of the way.

- String or small gauge rope – to tie up the placenta post-foaling.

- Sweat rug – for mare, post-foaling.

- Towels – to help dry foal.

- Navel treatment solution – to treat navel to prevent infection. Get this from your vet beforehand.

- Electrolytes – for mare to have in light feed, post-foaling.

- Feed – light feed such as a warm bran mash.

- Synthetic colostrum – in case the mare runs her milk (see page 54) and the colostrum is not collected.

- Bottle and teat – suitable to feed foal if it is weak and cannot get up to take its first feed.

- Disposable gloves.

- Surgical scissors – to be used if the mare's vulva has been previously stitched and the stitches need to be cut open for foaling.

- Spare set of suitable clothing.

- Wellington boots.

- Disinfectant – it is sensible to put a disinfectant bucket outside the foaling stable to be used by all people entering the stable.

- Two buckets – one for fresh water, the other for antibacterial wash.

- Bin liners – for collecting up rubbish and afterbirth.

- Foaling monitoring system – a variety of options are described below.

Monitoring systems

There is now a variety of foaling monitoring systems available on the market. Some are based on the mare's level of sweating and others on the mare's body positioning.

- The **sweat-sensitive foal alarm** consists of temperature sensors located on either side of the neck part of a breastplate, with a control panel to set the sensitivity. When the mare starts to sweat, a signal is sent to the transceiver (usually located by your bed!), which sets off the alarm. This system can be prone to false alarms, as the mare can get quite hot anyway in the latter stages of pregnancy, but without foaling being imminent. However, it does have a sensitivity panel so can be set differently for individual mares. It is also useful for monitoring a horse that has had a bout of colic, as any sweating will set off the alarm.

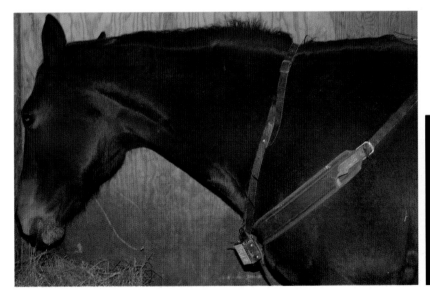

A mare wearing a sweat-sensitive foaling alarm.

The transmitter panel settings on the foaling alarm.

The sweat receptor that is positioned on the neck of the breastplate.

- The **position-sensitive foal alarm** consists of sensors on either the mare's headcollar or a roller. They are triggered by the mare going through a sequence of getting up and down and finally laying her head flat.

- **Baby monitors** are another useful system for listening in on your mare. However, they do require a trained ear; hay-munching sounds should be ignored!

- **Closed-circuit TV** is ideal for being able to monitor the mare without disturbing her. They are particularly useful for monitoring the mare with her newborn foal. However, for pre-foaling observation they do require the watcher to be awake! They are most useful in conjunction with a baby monitor.

Over the years, our stud has found that a combination of a baby monitor, sweat-triggered foaling alarm and closed-circuit TV provide the most effective and reliable monitoring system.

Foaling down at grass

For some mares, foaling down at grass is a more suitable option. The types of mare most likely to foal down at grass are native ponies and other pony mares due to foal in May/June, when the average temperature is warmer and the overall weather conditions are more favourable. The summer grass is usually in abundance, providing the mare with a good level of nutrition for milk production.

These pony mares rarely have problems foaling and need little close monitoring. However, it is important to locate and check the placenta carefully, and to bring the mare and new foal in to. check them thoroughly.

Native and pony mares usually have very strong natural instincts and their foals tend to have equally strong instincts to get up and suckle very quickly after birth. Be aware that some pony mares can be very protective of their newborn foals, especially if you have any dogs present.

Most mares will foal down in the morning (in the wild this gives the foal the best chance of survival by getting warmed and dried in the morning sun, enabling a quick recovery).

Mares can foal down in the field with other mares present, but they usually take themselves away from the group to foal. It is best if the mare can be in a field on her own, but if not the field must be big enough so that the mare can get away from her fieldmates.

CHAPTER SEVEN

Foaling down – stage by stage

Parturition – the process of giving birth

Stage one

- The first stage starts with the onset of uterine contractions, i.e. the mare goes into labour.

- Stage one usually lasts about fourteen hours, but it is often difficult to tell when the mare is in the early stages of labour.

- During the last four hours, your mare will start showing signs of discomfort, including sweating, groaning, stable walking, tail swishing, looking at her flank and general restlessness. She may start to run her milk.

- If your mare starts to run her milk at this stage it is important to collect it, as this is the foremilk or colostrum. The colostrum contains all the antibodies that the foal requires. Collect this milk in a bottle so that you can give it to the foal as soon as it is born.

- The first major contractions dilate the cervix and allow the outward bulging and rupture of the outer placental membrane (called the chorioallantois, CA) often referred to as the 'water-bag'.

  The presence of the CA and/or the feet of the foal in the vagina trigger the mare to start straining and initiate the onset of stage two.

Parturition/Labour Stages

Stage 1	Stage 2	Stage 3
Lasts about 14 hrs	Lasts 5–30 mins	30 mins–2 hrs after foaling
Mare sweats, gets up and down frequently	Water-bag breaks; delivery of foal	Passing of the afterbirth

Stage two

- The second stage of labour is brought on by the rupture of the water-bag (CA). The allantoic fluid (yellow urine-like fluid) flows out when the mare moves around and gets up and down.

- The mare's straining now increases in force and frequency. This second stage is fairly brief, ranging from five to thirty minutes. During

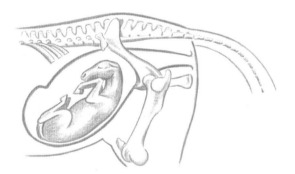

Before the first stage of labour.

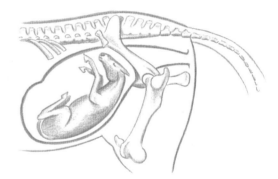

During the first stage.

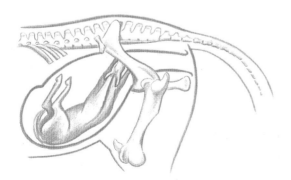

First stage, continued.

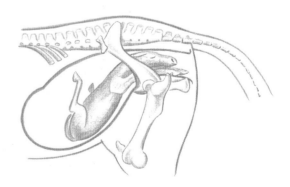

Early in the second stage.

The second stage. The foal begins to emerge.

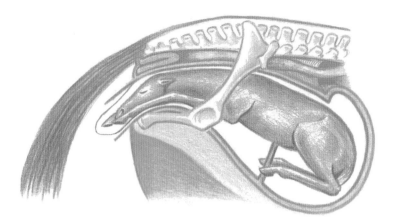

this time, the mare will get up and down repeatedly. She may well roll to assist in getting the foal into the birth position.

- Foals are born in the 'diver's position', with both forelegs extended and the head and neck following.

- One of the foal's feet usually appears at the vulva just after the rupture of the water-bag (CA). The second foot is often visible, about 10 cm (4 ins) behind the first, followed by the muzzle.

- The normal presentation of the foal's head into the vagina will cause the mare to lie down. Further straining occurs and the mare will usually lie flat out, with her legs straight, pushing for rapid passage of the foal.

- Once the feet, head, chest and hips are delivered, the mare will stop straining, leaving the foal's hind feet in her vagina.

A healthy foal just arrived and still attached via the umbilical cord.

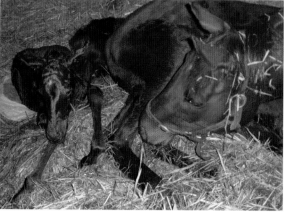

The newborn foal sitting up on its own.

- Usually, the allantoic membrane remains intact during delivery and surrounds the foal. It is important that the foal is encased in amnion (a white membrane) and not in the chorioallantois (a red velvety membrane). If a red membrane is visible, call your vet immediately. It may mean that the placenta has already come away, and this endangers the life of the foal.

- The foal should struggle to sit up and free the membrane from its muzzle. If, however, this doesn't happen, you should intervene to tear the membrane away to prevent suffocation.

- A normal foal should be able to sit up within five minutes of birth. If the foal cannot sit up unaided, assistance should be provided.

- The mare will usually lie resting for up to twenty minutes following delivery – this may vary from mare to mare. The struggling foal will break the umbilical cord, usually near to its abdomen.

- It is important that the mare does not get up too quickly at this stage and break the umbilical cord, as this can result in tearing the cord away from the foal. The cord should be allowed to 'stretch' gradually by the foal struggling, and break naturally after all the placental blood has flowed into the foal. It is at this point that the navel treatment solution should be poured liberally over the navel area to protect the foal from any infection setting in.

- A normal foal will have an immediate suckling reflex, and start making contact with the mare. The mare will often call to the foal and

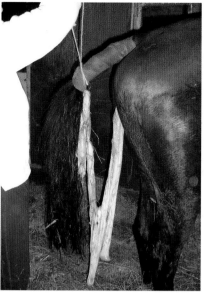

The remains of the umbilical cord are treated with a navel dip to prevent infection.

RIGHT: Tying the placenta to the tail with a piece of string to prevent the mare treading on it.

start to lick her foal clean. Once the cord has broken, the foal can be moved to the head end of the mare so that she can begin 'mothering' her foal. She will clean the foal and whinny to it.

Stage three

● This is the stage where the placenta (afterbirth) is passed, which can take anything from thirty minutes to two hours following foaling. During this time the mare will be up and the foal should have started nursing (suckling the udder for milk). To prevent the risk of the mare treading on any of the cord or amniotic bag that is already exposed, tie it up to the mare's tail to get it off the ground with a piece of string or rope.

● The foal may need some assistance in finding the teat to start drinking. It is vital that the foal starts drinking soon to get the all-

Normal foal activity post-birth

40 mins–1 hr	Up to 2 hrs	1–4 hrs	5–10 hrs
Foal should be standing	Foal should be suckling	First meconium should be passed	First urine passed

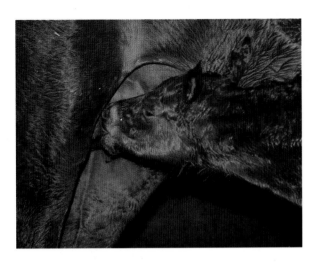

The newborn foal successfully suckling and receiving its first colostrum.

important colostrum (fore milk). The act of suckling also results in the mare producing oxytocin, which aids the release of the placenta.

- If the mare has not passed the placenta within four hours, veterinary assistance should be sought immediately. The vet will administer oxytocin to increase contractions and, it is hoped, release the placenta without further intervention.

- Once the afterbirth has been passed, it should be removed from the foaling area and laid out for examination. You need to make sure that none of the tissue is missing. If you think the afterbirth is incomplete, contact your vet and keep the afterbirth for him/her to view it.

Laying out the afterbirth so you can see that no tissue is missing.

59

Retained placenta can be very dangerous for mares, even if only a small amount of tissue is involved. Retained tissue can cause a major infection very quickly. If you suspect a problem, monitor your mare's temperature closely (a rise could indicate an infection). Your vet will probably attempt to remove any residual tissue and give the mare a course of antibiotics to ward off infection.

The importance of colostrum

- Foals are naturally born without any antibodies of their own. They receive them from their mother via colostrum, the first milk produced by the mare. As stated elsewhere, it is rich in antibodies that give the foal protection against infection.

- If the foal does not suckle within the first two to four hours of life, but you have been able to collect some colostrum from the mother, call your vet and he will administer the mare's colostrum via a stomach tube. If you have not been able to collect any colostrum from the mare, administer the synthetic colostrum that you should have in storage. (Synthetic foal colostrum is now available in paste form from specialist equine health companies, and it is a good idea to have some in stock in case it is required.) Your vet should be consulted to check the immune status of the foal to ascertain whether further treatment may be required. Early suckling of the foal is extremely important to aid the milk let-down of the mare and get the foal off to a strong start.

- If the foal has a good suckle reflex but has not been able to stand, then bottle-feed the foal with colostrum stripped from the mare.

- If your mare runs her milk before foaling and you have not been able to collect it, ask your vet to arrange a donor source of colostrum or an alternative source of antibodies, before the foal arrives.

- It is important that colostrum is received by the foal within the first twelve hours of life, as after this time the antibodies are broken down by the foal's digestive system rather than being directly absorbed.

CHAPTER EIGHT

Monitoring your mare and foal

The newborn foal

Once the foal is free from the umbilical cord it will struggle to get to its feet. This is a fairly precarious event as the foal's legs are incredibly long in comparison to its body and it finds them fairly awkward to handle at first.

The base of the foal's feet is soft, and the bottom of the foot does not resemble a normal 'sole' that you would expect. Do not be alarmed by this. This is to prevent the foal from damaging the inside of the mare, and this soft tissue disappears soon after the foal starts walking around.

- Assistance in keeping the foal on its feet may be required until it finds its balance. The foal will immediately start to search for the teat, often looking in all the wrong places! You can aid the foal by directing it to the mare's udder and teat. Once the foal has located the teat for the first time it rarely needs help again.

- Most foals will seek out the teat, latch on and drink within the first hour of standing up. However, some may not be successful in finding the teat to latch onto. Assistance can be given by getting the foal to suck on two fingers, and then (with your other hand around the foal's

The base of a newborn foal's foot, showing the protective soft tissue.

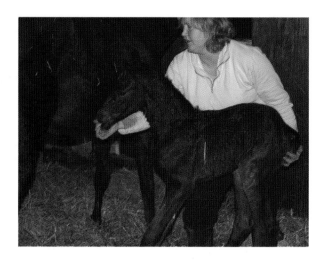

A newborn foal with a good suck reflex being encouraged towards the mare.

rear) encouraging the foal into position with its head under the flank of the mare reaching up to the udder. With the foal still sucking on your fingers gently manoeuvre the foal's muzzle onto the teat. Stay in position 'holding' the foal whilst it successfully drinks on its own. Once you know that the foal has successfully drunk from the mare and received the all-important colostrum, retreat and monitor the mare and foal from a distance, ideally via a closed-circuit TV or regular viewing through a window. Try and give the mare and foal some peace to be able to rest and acclimatise to each other.

- If your mare appears to be ticklish or squeals as the foal tries to suckle, it is sensible to ask someone to hold the mare while you help the foal to latch on. Once the foal has started suckling, the mare usually calms down and is happy to let the foal continue. In extreme cases when the foal is not being allowed to suckle, seek professional help.

Newborn foal passing meconium.

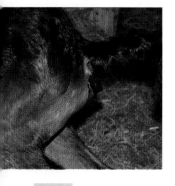

- The next important stage is the passing of meconium. This is the first bowel movement of the foal and is dark greeny/black in colour and sticky in texture. Meconium should be passed within the first few hours after birth. The foal may be seen to strain quite a bit when passing it. If the foal strains a lot, however, or rolls around and does not seem to be passing anything, call your vet immediately as there may be a blockage. Meconium retention is not that unusual and failing to deal with it promptly can be fatal.

- After the meconium has been passed and the foal is drinking well, the faeces will change to a milkier yellow colour.

- For the first twenty-four hours after birth, monitor the foal well to make sure it is drinking, and passing urine and faeces.

In no time at all, a healthy foal will be bouncing around the stable, in between periods of sleeping and feeding.

Normal behaviour to expect from a newborn foal

- The foal should be born with its muzzle clear of the placenta so that it can breathe for the first time as soon as it is born. It should be obvious from the movements of its nostrils and chest that it is breathing.

- The foal's eyes should be open and bright at birth.

- Immediately after birth, the membranes of the foal's mouth and tongue may appear quite a dark pink colour because of the normal pressures of the process of birth, but these should assume a normal light pink colour relatively quickly, once a normal breathing pattern has been established.

- Immediately after birth, the newborn foal's respiratory rate is rather high (60 breaths per minute) and its heart rate, which can be measured by putting a hand on its chest just behind the elbow, should be in the range of 80–100 beats per minute.

- Once the foal starts to recover from the stress of the birth process and take an interest in its surroundings, it should make attempts to rest on its hindquarters and sit up. This usually happens within a few minutes of being born. This actions aids respiration and indicates that the foal basically knows which way is up. It may make numerous attempts to stand. Most normal foals will stand within forty minutes to one hour of being born.

- Once standing with some confidence and stability, the foal should start to make attempts to suckle from just about anything it encounters. This might include the mare's elbows, nose, legs, the stable walls and you, if you're in the way. This behaviour

63

merely indicates that the foal is instinctively seeking out the udder. Most foals are suckling from the mare within hours after birth; call your veterinary surgeon if the foal has not had a good suck of milk by four to six hours of age.

- Apart from udder-seeking behaviour, newborn foals are not terribly inquisitive. Having discovered the udder, they tend to go back for frequent small feeds and are quickly able to stand up and lie down again for a rest or sleep, as and when they choose. The mare's teats should appear permanently wet or shiny, showing that the foal has been suckling, and the foal should lie down and sleep after suckling, indicating that it has been satisfied. The foal quickly establishes a 'rhythm' of frequent feeding and sleeping. In a normal foal any disturbance will quickly make it jump to its feet if it is lying down.

- Over the next twelve to twenty-four hours the foal will become increasingly interested in its surroundings and will have bonded closely with the mare – so much so that it will call her if she is not in immediate sight and will follow her if she moves or is moved from one place to another.

- The foal should appear bright and alert and will start to engage in periods of play, 'prancing' and 'chasing' around the mare between periods of feeding and sleeping.

- The foal should pass its first meconium from one to four hours after birth. Further meconium can be passed quite quickly or will follow over the next few days. The foal will urinate between five and ten hours after birth. Continuous urination is not normal – consult your vet.

- If your foal deviates from any of the expected 'normal' behaviour of a newborn then consult your vet.

The mare

Post-foaling, the mare may get cold after sweating during the birth process. Have a sweat-rug and a cooler-style rug available to aid drying and keep the mare warm. Many foaling boxes are fitted with infra-red heat lamps for warmth. If so, rugs will not be required.

Many mares are quite hungry after giving birth. A light feed, such as a warm bran mash with an electrolyte supplement, is a good idea. You can offer the mare a normal feed a few hours later. Have good quality ad-lib hay available at all times, likewise fresh clean water. To avoid the foal knocking over a water bucket in the first few hours, move any buckets out of harm's way and fix at a higher level, e.g. suspended from a tie-ring.

For most mares, motherhood comes naturally. Generally the new mother will stand quietly while the foal drinks, keeping her foal close at all times. However, a few mares find having their udders touched unbearable at first and will kick and squeal at the foal when it approaches. In this case, try to take some milk off the mare into a bottle until she gets over the initial dislike of having her udder touched. It is sensible to get some experienced help, as some mares can be very difficult at first.

Once the mare has passed the afterbirth and all seems well, she should

A day-old foal having a rest between feeds.

Mare with a two-day-old foal having a leg-stretch outdoors.

be watched over the next few days to see that she is eating well, looks good in her coat and is generally healthy. If her coat looks stary, she is off her feed or looks below par, then monitor her temperature and consult your vet if her temperature rises. This could be an indication that not all of the afterbirth has been passed and she is developing an internal infection.

CHAPTER NINE

The first few days

Most foals are born in early spring when the weather is not very reliable. Therefore it is usual to keep the mare and foal in their warm, dry foaling box for the first couple of days.

After that, when the foal is drinking well and strong on its feet, the mare and foal can enjoy a short period in the paddock to have a run around and light graze. Each day, providing the weather is suitable, you can increase the amount of time they spend out in the paddock.

From five to ten days post-birth, the mare will have her **'foal heat'** (i.e. come into season). In theory she can be bred on this heat, although the percentage of mares that take on their foal heat, compared with those on the first regular cycle that follows, is low. During the mare's foal heat, the foal may well **scour** (pass very runny yellow faeces).

The scour can sometimes 'burn' the skin on the foal's rump. Pasteing some Vaseline on the affected area as soon as the foal begins to scour can help prevent soreness. During this time, keep a close eye on the foal to make sure that it does not feel ill and reduce its suckling. Consult your vet if you find the foal beginning to look 'tucked up' (i.e. has its stomach drawn up).

Once the first ten to fourteen days are passed, the foal will be strong and running next to its mare. It will now be possible to integrate it with other mares and foals in the paddock. Make sure that the paddock is big

Foals produce milky faeces after passing meconium.This becomes runny and they 'scour' when the mare has her 'foal heat'.

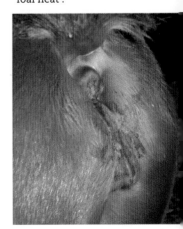

Mare and foal enjoying
summer grazing.

enough to allow the others to keep their distance until they are settled
with their new companions. Once happy, the mares will allow the foals
to play together and venture further away from their side.

CHAPTER TEN

Problems with newborn foals

If you suspect your newborn foal has a problem – act quickly. A newborn foal can deteriorate very rapidly. The sooner help is obtained, the better. Do not adopt the 'wait and see' approach – call your vet immediately.

Here are some of the problems that might be encountered:

Stupid or 'slow' foals

Some foals are 'slow', almost as if they have a headache. This may be the result of a traumatic birth or pressure on the head during delivery. Generally, these foals will need help to start suckling and orientate themselves to the mare, but they do improve rapidly.

Navel infection – joint ill

This is not typically 'newborn' but can occur within three days post-partum. It's an infection that occurs through the navel. The foal is most vulnerable at birth, which is why a navel dip or spray is recommended for use when the cord breaks. Infection that enters via the navel can cause septicaemia and joint ill (infection of the joints).

Joint ill becomes apparent at the end of the first week or as late as three weeks after birth. The foal may be off-colour for a few days before the full symptoms appear, which are lameness involving one or more limbs, a high temperature, swelling and heat in one or more joints. The swelling may move from joint to joint.

Joint ill is very serious and has a high mortality rate. Early detection and immediate veterinary treatment are vital if there is to be any chance of a full recovery.

Limb deformities
If severe, these may render the foal unable to stand. A vet should be sought immediately as many abnormalities can be treated successfully.

Brain damage
A perfectly normal-looking foal may suffer from brain damage, caused by the lack of oxygen at birth. The foal may exhibit no suck reflex and be unaware of its surroundings. In severe cases, the foal may convulse.

Infection
In some cases the foal may have developed an infection in the later stage of pregnancy and may be born too weak to function normally.

BELOW: Newborn foal with signs of contracted tendons. RIGHT: The same foal twenty-four hours after treatment.

Contracted tendons

This is apparent when one or more legs appear very 'over at the knee' or excessively straight at the fetlock. There are a variety of treatments for this condition, all of which require the vet. Speed is very much of the essence when treating this condition. In many cases it is possible to get a good result, leading to a foal with normal limbs.

Meconium retention

Signs include straining, squatting, lifting the tail, rolling, looking around at flanks, lying on back and general colic-like symptoms.

Ruptured bladder

Signs of a ruptured bladder can appear two to three days after birth. The symptoms are similar to meconium retention, but in addition the abdomen becomes distended and only small amounts of urine are passed. This condition requires surgery.

Undershot jaw/parrot mouth

This is an inherited feature. In some cases it can be corrected by surgery, but the best cure is not to breed from any mare or stallion exhibiting such a condition.

Umbilical hernia

This is an extended swelling which appears at the umbilicus (where the umbilical cord was attached) as a result of a gap in the abdomen muscle, at about six weeks of age. The swelling steadily increases in size, depending on the size of the gap in the muscle. Most disappear over time, but large ones may require simple surgery to close the gap. This can be an inherited condition.

CHAPTER ELEVEN

Your mare and foal until weaning

If the foal is born early in the year, such as February or March when the weather can vary, it is sensible for your mare and foal to be stabled at night and for the mare to have two hard feeds a day. The grazing value is poor at this time of the year and the lactating mare will require a high plane of nutrition to maintain a good level of milk production.

For foals born in late spring, when the days are longer and the quality of grazing is better, your mare and foal may be able to stay out at night.

How you manage the grazing of your mare and foal will often depend on the breed type. For example, a Thoroughbred type mare may well need a good level of nutrition to maintain herself and her milk supply. So, regardless of the time of year of foaling, she will require two hard feeds a day and stabling at night. A native pony type mare, however, who normally has to be closely monitored for laminitis, will require little or no supplementary feed and can be kept out at grass full time.

As the foal develops, its diet will gradually change from milk only to milk plus fibre, and then mostly hard feed and fibre with a small amount of suckling.

From as early as a few days old, the foal will start grazing when out with its mother, but at this stage it will be reliant on its mother's milk as its main source of nutrition.

A five-day-old foal nibbling at grass while resting.

From as early as three weeks old, you will notice the foal start picking at its mother's hard feed. When it gets to the point when it is trying to share its mother's feed regularly, introduce another feed bowl, placing it on the ground for the foal, to reduce the likelihood of the mother getting fractious.

As the foal approaches six months of age it will have become very independent of its mother, spending much of the day grazing (with other foal friends, if any) and only suckling small amounts. The mare will often begin to start to self-wean the foal by not always allowing it to suckle.

At this stage, or anytime from five months of age onwards, the foal can be weaned from the mare.

Training your foal to lead

Like all animals, it is much easier to get foals used to new things when they are very young, rather than leave it until they are older and more independent-minded. It is therefore sensible to introduce your foal to wearing a headcollar and learning to lead at an early stage when little resistance will be encountered.

It is a good idea to get the foal used to being well handled in the first few days while it is still indoors. Stroke it all over its body, including behind the elbow (where the girth will go in future) and the ears, around its rump and along its tummy. To start with, your foal will find it ticklish but it will get used to it very quickly and actually begin to like it. You can do this morning and evening while your mare is eating her food, as she will be settled.

- Always use a leather headcollar for foals. It is far better that the headcollar breaks if the foal gets caught up, rather than the foal hurt itself with an unbreakable nylon headcollar. The latter may be cheaper but could easily prove a false economy. Foal headcollars come in a variety of different sizes to accommodate the foal as it grows, starting with a foal slip.

- Start teaching your foal to lead as soon as possible. With one arm around the foal's rump and the other around the chest, encourage the foal to walk forwards. Then progress to having one arm around the

A five-day-old foal having its first leading lesson.

The mare is led slightly in front of the foal to encourage the foal to lead.

foal's rump and the other on the headcollar and, without pulling on its head, start walking forwards.

- When teaching your foal to lead have an assistant leading the mare just ahead of the foal so that it will naturally want to move forwards, following its mother.

- As the foal gets more confident and is happy to walk forwards by being led on the headcollar, you can progress to using a lead rope.

- It is good practice to lead your mare and foal out to the paddock every day and to remove the headcollar at night and replace it in the morning. This is more difficult if your mare and foal are living out, but it's worth making the time to have regular leading lessons.

- When leading your foal always position it next to the mare, usually just behind the mare's shoulder, on the mare's nearside. The foal will naturally follow the mare alongside by the flank, so using this position will aid the learning process.

Foal farriery

Rather like teaching your foal to lead, the earlier you start handling the foal's feet, the easier trimming and other farriery work will be.

Out in the field, enjoying
a rest.

- It is advisable to start to pick up your foal's feet as soon as it is used to being held. Always have two people: one to hold the foal and the other to pick up the feet.

- You should introduce the farrier at about six weeks of age. At this point the farrier will probably only want to run the rasp around the foal's feet and examine the foot shape. It is important that the foal is stood up (next to its mother) on a hard, flat surface so that the farrier can view the foal's foot shape properly.

- If the foal's foot shape is poor, or the feet are not a pair (front or back feet) a farrier can aid the long-term foot development very dramatically through regular remedial work from an early stage.

- The foal should be trimmed regularly every six weeks from its first farriery session. You will need to keep an eye on the foal's feet as they go through periods of growth spurts, and it may be that they need to be trimmed at shorter intervals, such as four weeks. This is particularly the case from six to twelve months of age.

- If you are concerned about the shape or angle of your foal's feet, consult your vet and farrier about remedial work.

Vaccinating your foal

If your mare was properly vaccinated prior to foaling, your foal will have cover against tetanus toxoid and equine influenza until four months of age. At four months of age you should arrange with your vet for your foal to have its first tetanus toxoid and equine influenza injection. Your foal will require three injections to set up proper immunity; one at four months of age, one at four to six weeks after the first injection, then a third at six months after the second injection. From then on your foal will require an annual booster to maintain cover. All injections must be recorded in the foal's passport.

Registering your foal

It is now a requirement that all equines have a passport. If you have used a registered stallion you will have been provided with a covering certificate when collecting your mare from stud. On the covering certificate is a section for foal registration once born. It is usually a requirement that either a vet or an authorised person (authorised in the UK by the

A well-handled three-month-old foal being presented for evaluation. This foal was awarded a Premium.

Passport Issuing Organisation, PIO) completes the foal identification certificate.

Once the certificate is complete, send it with the required registration fee to the relevant passport issuing authority. You will usually receive your foal's passport within about three months. All vaccinations must be recorded in this passport, which must always be held at the same location as the horse.

Worming your foal

During the gestation period, the foal acquires no antibodies to protect it from worms, and this makes the foal susceptible to infection. There are two species of parasite that are of particular concern: intestinal threadworm and large roundworm. Over time, horses develop a natural immunity to these species of parasites.

Intestinal threadworms

Infecting the foals soon after birth, usually via the mare's milk, these small worms – up to 1 cm ($1/2$ in) in length – live in the small intestine of the foal. Because the lifecycle of this parasite is very short, foals as young as four weeks of age can develop heavy worm infestations, causing diarrhoea, loss of appetite and dullness. However, natural immunity is acquired at around six months of age.

Large roundworms (Ascarids)

Practically indestructible, the eggs of the large roundworm that contain the infective larvae are surrounded by a thick, sticky outer coat. No matter how clean the stables, these parasites eggs will be present. These eggs can survive for years on items such as buckets, walls, bedding and pasture, and even on the mare's udder. The main source of infection is eggs in manure: all foals eat fresh manure every few hours in order to 'seed' their digestive tract with the beneficial micro-oganisms essential for the proper digestion of vegetable matter. Once ingested, the eggs hatch and the larvae of this parasite migrate through the foal's blood-stream to its liver and lungs, causing coughing, fever and loss of appetite, before they return to the small intestine. In the small intestine they develop into adults. These adults reach up to 40 cm (16 ins) in length, and can be present in the foal's small intestine from twelve weeks of age. Heavy burdens of adult worms cause weight loss and can, in rare cases, result in a rupture of the gut. Natural immunity is acquired at around eighteen months of age.

Damage inflicted to the foal by parasites may be gradual and subtle, with no noticeable signs from the onset. However, the foal's growth and development will become impaired, its performance ability reduced, and its resistance to disease lowered – resulting in potential complications in the longer term. To protect the foal and give it a head start, worming is vital.

When to worm

The worming of foals should start at four to six weeks of age, or as soon as they start to graze on pasture, with treatment being repeated every four weeks until six months old. Thereafter routine worming can be continued as per wormer manufacturers' recommended dosing intervals. With regard to tapeworm, infestation is unlikely to occur in foals under two months of age, so treatment of foals below this age is not considered necessary. Tapeworm is controlled by means of a double dose of a pyrantel-based wormer or with a single dose of a praziquantel-based wormer.

WORMING TIPS

- When choosing a wormer for a foal it is important to avoid drugs not licensed for use in very young foals.

- Dose the foal according to its body weight. Under-dosing results in more worms surviving, leading to pasture contamination, as well as contributing to the development of resistance. Overdosing can cause undesirable side-effects. As the foal is developing, its weight will be constantly changing; therefore, foals should be weighed prior to each treatment, either by means of a weigh tape or ideally weigh scales.

- A paddock rotation system should be adopted so that nursing mares and their foals do not graze the same area in successive years. Ideally rotate the land with sheep or cattle, as worms that affect horses are host specific and cannot survive in sheep or cattle, so any larvae they eat will be destroyed.

- Avoid turning out youngstock in small turn-out paddocks as pasture will develop extremely high larval counts, particularly if droppings are not removed each day.

By adopting a rigorous worming programme, supported by pasture management, the risk of parasite damage to the foal can be reduced, thereby giving the foal a good start so that it develops into a beautiful, healthy horse.

CHAPTER TWELVE

Weaning your foal

Weaning can be a traumatic time, but if well managed there need be very little impact on mare and foal.

Weaning procedure on studs

At well-established studs, weaning is an annual event that causes little distress to the mares and foals involved. Usually a group of mares and foals that have been grazing together for a while will all be weaned together. Often a 'nanny' mare will be used. A 'nanny' can be an older mare (probably retired from breeding) that has a quiet nature. She will have been grazing with the group of horses prior to weaning so that the foals are familiar with her. This 'nanny' mare will provide stability for the foals once weaned (separated from their mothers).

The foals

It is sensible to wean foals mid-morning after their morning feed. The mares are led away all together to another yard that is some distance from the foals, so that they cannot hear each other whinnying nor see each other.

The foals should remain in the stables that they are used to, with the top door shut to reduce the chance of any of the foals trying to jump out. At

A healthy foal.

least one person should remain in the foal yard to monitor them. It is a good idea to remove any buckets, loose mangers etc. from the stables prior to weaning to reduce the risk of unnecessary injuries.

The ideal weaning yard is a group of 'Loddon' style boxes that allow the foals to see and 'talk' to each other through the bars.

The foals will be upset, but usually they will all have settled down within an hour or two and you can replace water buckets and provide a small feed.

The foals can then go out the next day with the nanny mare.

The mares

When the mares are moved to another yard you will notice that they are usually quite happy to leave. By six months most mares will have been naturally weaning their foals. Many mares will already be in foal again so the drain of being pregnant and having a foal at foot is quite demanding on their body.

When the mares are first taken away from their foals, it is sensible to put the mares out in a field. To begin with they usually run around the field whinnying, and this will start their milk running.

After approximately ten days, your mare should be successfully weaned and her udder not producing any more milk. The mares and foals must be kept separated for at least six months.

Weaning a single mare and foal

Weaning a single mare and foal in a small area can be difficult and I would advise sending them to a bigger stud to join in a group for weaning. Take them to the stud a couple of weeks before weaning to allow them to settle in. Then they can be weaned along with the other inmates, which is far less traumatic.

If you are unable to use a stud then start your weaning process by separating the mare and foal at night. Stable the foal next to mare at night to reduce suckling. This will start a gradual reduction in the mare's milk production and get the foal used to being on its own with its mother close by. Introduce a companion, which at first can go out with the mare and foal in the paddock so that the foal is familiar with its new friend.

Once the foal has got used to being separated at night, remove the mare to another location and put the companion in the mare's stable next to the foal. The foal should be comforted by the presence of the companion.

Make sure that the mare is far enough away not to be able to hear/see or be heard/seen by the foal.

Drying up the milk supply

Once removed from her foal, the mare's milk supply takes about ten days to dry up naturally. During this time, she will 'run' her milk – if she does not, she may develop **mastitis**, so it is important that she is monitored and managed well during this period.

The mare's milk will run as she moves around, and over the next few days her milk production will reduce due to the lack of suckling. It is very important that this is done naturally and that the mare is not actively milked by hand. Milking will encourage the mare to carry on producing milk, as it will be perceived by the mare's system as suckling. So a sharp eye should be kept on the mare's udder to make sure that there is evidence of her running her milk (look for dried milk on the inside of her legs) and that her udders have not become overly swollen. (Please note that for the first few days the udders **WILL** appear larger.)

You can aid the reduction in milk production by firstly reducing the mare's feed intake during this 'drying up' period, and secondly by keeping her out at grass so that she is moving around rather than standing in a stable.

- The first signs of mastitis are the mare looking off-colour, a stary coat and a high temperature. If affected, her udders will become hard to the touch (and very painful, so she may well kick if her udders are felt), and the milk will appear lumpy (if you attempt to milk her just to check the consistency). If you are at all concerned take your mare's temperature regularly; if it is raised then call the vet straight away. Do not delay in calling the vet, as mastitis can get worse very quickly once it starts to take hold.

Appendix 1

Pregnancy checklist and progress chart

MARE	At least a month prior to going to stud	4–5 days prior to season (oestrus)	1st day of season	End of season
	Vet to take swabs to have certificates prior to going to the stallion	Take mare to stud or AI centre	From now on mare will be examined for ovulation by scanning and/or teased by stallion until mare is receptive to the stallion for mating	Mare not receptive to teasing stallion. She has now 'gone off'
	Check vaccinations are all in order			

16 days from ovulation (or first covering date)	24–30 days from ovulation (or first covering date)	40 days from ovulation (or first covering date)	60 days from ovulation (or first covering date)	0–100 days from ovulation (or first covering date)
Mare scanned for pregnancy	Mare scanned to ensure that pregnancy has developed well	Earliest date of pregnancy diagnosis by blood-test	Second blood-test to confirm continuation of pregnancy	Supplementary progesterone may have been prescribed for older mares who have a poor record of maintaining pregnancy
	Mare can go home if healthy pregnancy	Last chance to abort if mare carrying twins		

Appendix 2

Mare and foal care: late pregnancy to weaning

	3–6 weeks before due date	4–5 days before due date	24–48 hrs before foaling	Foaling date
MARE	Tetanus vaccination to cover mare and foal Approx 4 weeks prior to due date, the udder will develop	Worm prior to foaling Mare's back end becomes floppy	Wax may be seen like a cork on the teats of the mare's udder	Colostrum may be expressed from the mare
FOAL				Vet examines new foal at 24 hrs unless emergency

0–6 weeks	7–12 weeks	13–19 weeks	20–24 weeks	Weaning 5–6 months
				Remove mare to wean. Monitor daily for mastitis until 'dried up'
4-6 weeks first worming	10 weeks second worming	14 weeks third worming (can now worm for tapeworm with Pyrantel-based wormer)	First tetanus toxoid vaccination at 4 months	

Appendix 3

Vital statistics

Gestation period of the mare	340 days +/-
Foal temperature (normal)	37.2°C–38.3°C
Foal temperature (at birth)	37°C
Foal respiration rate (normal)	80/min
Foal respiration rate (at birth)	60/min
Foal heart rate (normal)	80/min
Foal heart rate (at birth)	80–100/min

Glossary

AFTERBIRTH – the placenta.

AI – artificial insemination.

ANOESTRUS – period when mare is not receptive to the stallion for breeding during the winter months, as a result of short daylengths.

ANTIBODY – protective protein produce by the body in response to the presence of a virus or bacteria.

BREEDING SEASON – overall period when mare is receptive to stallion, usually from February to September (northern hemisphere).

CA – chorioallantois: section of the placenta that is joined to the uterus.

CERVIX – neck of the uterus opening into the vagina.

CLEANSING – the term used to describe the passing of the placenta.

CLITORIS – a body of tissue found just inside the vulva.

COLOSTRUM – foremilk, which includes antibodies.

DIOESTRUS – period when mare is not receptive to the stallion during the breeding season.

ECG – equine chorionic gonadotrophin (pregnancy hormone).

ENDOMETRIUM – tissue that forms the lining inside the uterus.

FOETUS – unborn foal.

GENITALIA – genitals (i.e. reproductive organs).

GESTATION PERIOD – pregnancy

MAIDEN MARE – a mare that has not been bred from before.

OESTRUS – period when mare is receptive to the stallion.

OVUM/OVA – egg/eggs.

PARASITE – an organism living off another organism, e.g. worm.

PARTUITION – giving birth; pre-partuition being before birth, and post-partuition being post-birth.

POST-PARTUM – period after giving birth.

SEASON – term used to describe the breeding cycle of the mare.

TEASER – stallion used to tease mare to find out when she is receptive to mating.

URETHRA – tube through which urine is discharged from the bladder.

UTERUS – womb.

UTERINE HORN – the uterus has two horns, the right and the left.

VENEREAL DISEASE – a sexually transmitted disease.

VULVA – external opening of the vagina.

Further reading

Infectious Diseases of Horse Order – Reference: 1987 No.790. Obtainable from HMSO

Equine Viral Arteritis Order – Reference: 1995 No.1755. Obtainable from HMSO

BEVA Code of Practice for the Use of Artificial Insemination in Horse Breeding. Obtainable from the British Equine Veterinary Association, Wakefield House, 46 High Street, Sawston, Cambridge,. CB2 4BG

The Genetics of the Horse by A. T. Bowling and A. Ruvinsky, CABI Publishing (2000)

Horse Genetics, Ann T. Bowling, CABI Publishing (1996)

Equine Color Genetics by D. Phillip Sponenberg, Iowa State University Press (2003)

Horse Color Explained by Jeanette Gower, Trafalgar Square Publishing (1999)

Modern Horse Breeding – A Guide for Owners, Susan McBane, Lyons Press (2001)

Equine Reproductive Physiology, Breeding and Stud Management, Mina C.G. Davies Morel, CABI Publishing (2002)

Horse Breeding, Peter Rossdale, David & Charles (2003)

Equine Stud Management: A Textbook for Students, J.A. Allen & Co.Ltd, (1992)

Equine Nutrition and Feeding, David Frape, Blackwell Publishing (2004)

From Foal to Full Grown, Janet Lorch, David & Charles (1997)

The BHS Veterinary Manual, P. Stewart Hastie, Kenilworth Press (2001)

The Complete Book of Foaling, Karen E.N. Hayes, Howell Reference Books (1993)

Index